DOGS

DOGS

TOM JACKSON

amber
BOOKS

This pocket edition first published in 2023

First published in a hardback edition in 2022

Copyright © 2023 Amber Books Ltd

Published by
Amber Books Ltd
United House
London N7 9DP
United Kingdom
www.amberbooks.co.uk
Instagram: amberbooksltd
Pinterest: amberbooksltd
Twitter: @amberbooks

ISBN: 978-1-83886-258-9

Project Editor: Michael Spilling
Designer: Keren Harragan
Picture Research: Terry Forshaw

Printed in China

Contents

Introduction

It is a well-worn phrase that a dog is a man's best friend, and a woman's, too. There is also no doubt that there is no other animal with which we share such a long and mutually fruitful history. What there is some debate about is how far back that story goes. Dogs and people have lived together for at least 15,000 years, but the duration of the process that led us to domesticate dogs could easily be double that. And did we domesticate them anyway, or did they tame us? Of course, today, the pampered pooches

across the world – fed and sheltered by loving owners – seem to be doing very well out of this owner–pet relationship! Something similar might have brought us together all those years ago. Packs of wild dogs were attracted to campfires knowing they could find scraps of meat and warmth. Over the generations – or perhaps faster – the human and dog families learned to tolerate each other, then to trust and then to become firm friends. Dog families integrated into ours, and we took control of their breeding, step by step developing the 360 or so breeds that we see today. Each breed has its own story, so let's take a journey to examine the life, work and play of dogs.

ABOVE:
Whatever this trio of French bulldogs is thinking, they aren't saying.

OPPOSITE:
A Lhasa Apso puppy, a Tibetan breed, is looking for something
– or someone – to play with.

Wild Dogs: the First Breeds

There have been dogs living somewhere on Earth for 40 millions years, around 20 times longer that we humans have been around. Before we came along, the dogs were well placed to take the title for the most widespread large mammals on Earth. They were – and still are – found on all continents excepting Antarctica, not just surviving but thriving in all climate regions, from the tundra of the High Arctic to the most arid of deserts. If it is anywhere that dogs succumb to competition from other large mammal carnivores, notably the cats, it is in the tropical rainforests, but look for them and you will find some species living even there.

Dog species form the *Canidae* family, which has an obvious division into two tribes: the dogs and the foxes. All share the dog body form, which is characterized by a large head with a wide jaw sitting on a slender body and supported by (usually) long legs. The jaw is the dog's primary weapon, used to kill and eat its prey. The body and legs are a supremely efficient transport system that allows dogs to travel for long hours at near maximum running speeds. The broad ribcage houses the engine of big lungs and a powerful heart, while the legs provide long, graceful strides.

There are 34 species of canid, going by such common names as coyote, dhole, jackal and culpeo, but the one with which we are most familiar is the grey wolf, *Canis lupus*. This large howling hunter, much feared in folklore, is the wild relative of our pet breeds.

OPPOSITE:
African wild dog
The scientific name for this rangy, big-eared animal is
Lycaon pictus, which refers to the blotchy fur pattern
that resembles spots of colour on a painter's palette.

Pack hunter

African wild dogs, seen here being forced to flee by a belligerent buffalo cow, live in large packs of 20 or more. While wolves are the main pack-hunting dogs along the fringes of North Africa and the world beyond, this species is the largest dog in the rest of Africa, albeit these days sadly classified as highly endangered.

OPPOSITE:

African wolf

A recent analysis of the golden jackals of Africa and Western Asia led to the discovery that the African animals are in fact a separate species of small wolf, now called the African wolf. Golden jackals are regarded as a wholly Asian and European species.

LEFT TOP:

Arctic fox

Equipped with a thick coat of white fur, the Arctic fox is able to survive year round in the icy land of the Arctic. It has small ears and shorter legs than other dogs, to reduce heat loss at the extremities.

LEFT BOTTOM:

Summer clothing

For much of the year, the Arctic fox has white fur to help it blend in with its surroundings. However, during the short summer, it develops a grey-brown coat.

LEFT:
Bat-eared fox
This small vulpine, or fox-like species, lives in the semi-deserts of southern Africa. The immense ears serve as both a radiator for shedding excess heat and acute sensors for tracking the movements of small insect prey.

ABOVE:
Bengal fox
A silver-furred fox living across South Asia, this species is typical of Old World foxes, in that it lives in small family groups but generally hunts alone.

OPPOSITE:

Black-backed jackal
A medium-sized African dog, which has catholic tastes, it feeds on carrion, insects and will tackle larger prey, often in pairs.

LEFT ABOVE:

Pack animals
These jackals of East Africa live in large packs. They work together to find food and defend it from larger animals, such as lions.

LEFT BELOW:

Bush dog
Looking rather different to its relatives, this small, short-legged dog is one of two species to live in the tropical rainforests of South America. The other is the short-eared dog, which dominates the western region of the continent.

Cape fox

This is a red fox species
that lives in the arid semi-
deserts of south-western
Africa. It eats insects
in the main, plus fruits
when those are available.

RIGHT TOP:

Corsac fox
These pale-coloured foxes live in the dry grasslands and deserts of Central Asia. They feed on rodents and insects.

RIGHT MIDDLE:

Coyote eyes
The coyote is the most widespread wild dog species in North America. It has learned to live in suburban areas, and has a reputation for its cunning.

RIGHT BOTTOM:

Lone howl
Although it lives and hunts alone, the coyote likes to let others know where it is. The message is: stay out of my territory.

OPPOSITE:

Coyote
Tracking its main prey with an acute sense of smell, coyotes prey mostly on small rodents. They are even known to team up with American badgers to find and then dig out ground squirrels and other burrowing prey.

RIGHT TOP:
Crab-eating fox
This South American fox hunts mostly insects in the dry seasons, but when rains turn their habitats to swamp, they go looking for crabs and other aquatic prey.

RIGHT BOTTOM:
Culpeo
Also known as the Andean zorro, this hunter patrols the arid slopes of the Andes and coastal strip of western South America.

OPPOSITE:
Darwin's fox
Named after Charles Darwin, who first identified this species during his famed voyage aboard HMS *Beagle* in the 1830s, this is an extremely rare species that's found only in a few patches of Chilean coastline and some offshore islands.

LEFT:
Fennec fox
One of two dog species to live full time along the edge of the Sahara Desert, the fennec fox is well known for its tall ears. These are filled with blood vessels that radiate body heat to help the animal keep cool.

ABOVE TOP:
Dhole
Also known as the Indian red dog, the dhole is a lupine species – in other words, a relative of the wolf – that lives in the forests of Asia. The small wild dogs are highly effective pack hunters.

ABOVE BOTTOM:
Ethiopian wolf
Found only in the Ethiopian Highlands, this endangered species lives in social groups, but hunts alone. It is a specialist predator of mole rats.

ABOVE:
Grey fox
Living alongside the red fox in North America, despite
the name, this dog species also shows red and brown
colouring as well as silver-grey fur.

RIGHT:
Gray wolf
This is the largest and most widespread of all dog
species. It grows to 1.6 m (5.25 ft) long and can weigh
80 kg (176 lb). The largest wolves live in colder areas.
The species is divided into many subspecies, such as
the timber wolf and tundra wolf.

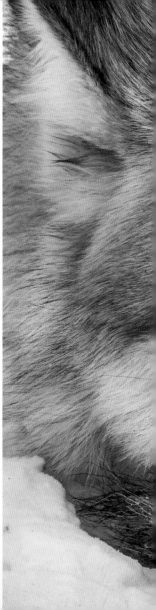

OPPOSITE:
Island fox
This is a relative of the grey fox that lives only on the Channel Islands off the coast of California.

LEFT TOP:
Maned wolf
Despite the name, this South American dog is actually more like a fox than a wolf, albeit with exceptionally long legs. Half of its diet is plant foods.

LEFT MIDDLE:
Blanford's fox
This species of fox lives in the arid deserts and hills of the Middle East.

LEFT BOTTOM:
Kit fox
One of the smallest species of fox, this lives in the deserts of the American Southwest.

LEFT TOP:

Rüppell's fox
This insect-eating species lives across North Africa and the Middle East.

OPPOSITE BOTTOM:

Pampas fox
This small canine species is named after the grasslands of South America where it hunts at night for small animal prey.

LEFT BOTTOM:

Sechuran fox
This South American fox is highly adapted to live in the parched Sechura region on the coast of northern Peru.

Red fox
This is the most widespread fox species. It lives in North America, Europe and northern Asia and has adapted to survive well in both urban and agricultural habitats.

LEFT:
Side-striped jackal
The largest jackal species in Africa, this pack canine lives in the south of the continent. It may hunt as a pack but is more likely to set out alone to find small prey.

ABOVE:
Raccoon dog
Looking very like the entirely unrelated raccoons of North America, this canine species lives in Siberia.

ABOVE:
Tibetan sand fox
Equipped with thick fur against the cold, this species lives on the high plateau north of the Himalayan great divide.

RIGHT:
South American grey fox
Also known as the chilla or zorro, this species lives in the cold, barren habitats of Patagonia in the far south of South America.

Hunting Dogs

Many of the most popular dog breeds were created over the centuries to help their human masters with hunting. Today, few people actually hunt animals for food. Instead, dogs are kept by professional sports hunters, who kill "game" for fun. But, of course, these breeds once used in hunts are now much-loved and affectionate pets as well.

Breeds of hunting dogs are not easily delineated from other working dogs. For example, tough fighting breeds developed as watchdogs and for livestock protection are also trusty and perhaps essential companions for human hunters far from home. Nevertheless there are perhaps three roles that dog breeds fulfil: finding quarry, catching quarry and retrieving kills.

The first of these, finding game, is the job of scent hounds, which pick up the smells of a target and follow it – literally doggedly. Wild dogs will find prey in much the same way. Once prey is within striking distance, a sight hound could be used to chase and kill the prey. The breeder's skills ensure that the dogs do not eat the quarry, something a wild dog would do straight away. Smaller hunting dogs, such as terriers, are used to chase prey into their burrows should the need arise.

Gundogs are breeds that aid hunters armed with shotguns, and have a less blood-thirsty profile. Pointers and setters indicate game, spaniels flush it out, and retrievers find and bring the dead game – most often birds – to the shooting party.

OPPOSITE:
Bloodhound
The archetypal scent hound, the bloodhound has 200 million scent receptors in its nose and an acute sense of smell that's 40 times more powerful than ours.

OPPOSITE:

Hunting desire

A bloodhound with a pheasant. The dog has a high motivation to find prey and is hard to stop once it has a scent. This is harnessed by people for all kinds of hunts – not least ones to find fugitives from the law in North America.

ABOVE:

Basenji

This African hunting dog is hard to train but is a persistent and intelligent hunter.

OPPOSITE:
Ibizan hound
A popular fast-running Catalan species that is used to hunt rabbits by sight, sound and scent.

ABOVE TOP:
Airedale terrier
The largest of the terrier breeds, hailing from Yorkshire, England. As with all terriers, it once worked killing rats and other vermin, but in this particular case it was developed specifically to kill otters.

ABOVE BOTTOM:
Basset hound
Much loved for its slow, bumbling demeanour, this scent hound was bred to have short legs so it couldn't outpace its human handler as it tracked quarry.

Springer spaniel

Loyal and eager to please, spaniels have the job of flushing game birds into the air and into sight of the guns. It will then sniff out fallen birds and bring them back.

English foxhound

A breed that is smaller and faster than bloodhounds but still drive to follow a scent to its source.

Gentle temperament

Packs of foxhounds lead hunters on horseback in traditional hunts. Despite being used by huntsmen to track quarry, foxhounds generally display a gentle and affectionate temperament.

PREVIOUS PAGE:
German short-haired pointer
Intelligent and obedient, this breed, like all pointers, will freeze on the spot when it finds quarry. Its muzzle points directly at its target, indicating its location.

OPPOSITE TOP:
Italian greyhound
Greyhounds are sight hounds that are able to track fast-running prey by sight – and follow at high speed. This makes these breeds ideal for racing.

OPPOSITE BOTTOM:
Norwegian elkhound
This breed, which is well adapted to cold weather, was bred to track deer in winter forests.

LEFT:
Italian spinone
This is a pointer-type hunting dog from northern Italy. The breed can be traced back to the 15th century.

Japanese Akita
A big and capable
hunting dog named after
a region of Honshu, the
largest Japanese island.
It was bred to attack deer
and wild boar, and to
defend against bears.

OPPOSITE:
Pharaoh hound
Originally from Malta, this sight hound was bred to catch rabbits. In Maltese, its name translates as "rabbit dog", and its English name alludes to a debunked claim that it is a descendent of dogs from ancient Egypt.

LEFT & BELOW:
Dalmatian
This impressive pointer breed from Daltmatia, on the coast of what is now Croatia, is famed for its black spots on a white coat. While descended from hunting breeds, the Dalmatian was bred as a carriage dog, and ran alongside the conveyances as a symbol of the wealth and status of the passengers inside.

Irish wolfhound
As its name suggests, this sight hound breed was developed to go out and kill wolves that threatened livestock. It was up to the task, being one of the largest dog breeds of all.

Otterhound
This English breed had the job of killing otters that threatened the fish stocks in country rivers. The breed is in rapid decline today.

Labrador retriever
One of the most popular pet breeds in the world, this gundog was bred as a retriever. Despite the name, this breed is originally from the Canadian province of Newfoundland, not Labrador. In maritime Canada, it was originally used to haul in fishing nets and collect the fish that fell out.

Working Dogs

Since humans and dogs first began to live in close proximity with each other, dogs have been making themselves useful. The ancestors of today's domestic dogs were highly social wolves that lived and worked as packs, and so it was a small transition for a tamed dog to join a different team (the human one).

Perhaps the first job carried out by our canine assistants was guard duty. This was a symbiotic affair: the humans provided warmth, shelter and some food and in return the dog would do what came naturally – challenging any large animal, dog or human that came into the group's territory. The dog's superlative senses of smell and hearing meant that behaviour manifested as an early warning system – and also a defensive force that was instinctively aggressive to intruders. Today's guard dog breeds maintain that urge to protect the pack.

Guardian dogs do the same to protect domestic herds and flocks – very often from wolves. However, their close colleague, the sheepdog, harnesses another primordial instinct: to chase prey animals. In addition, working dogs are used by emergency services, the sports industry and to help the infirm and disabled. These dog jobs rely on other canine characteristics, such as a keen sense of smell, great stamina and considerable intelligence.

OPPOSITE:
Border collie
Radiating intelligence, this is the superlative sheepdog breed. The "border" in the name refers to the moorland region – filled with sheep farms – that forms the borderland between England and Scotland.

Busy breed
Border collies are extremely active and want
nothing more than to work, work, work. Despite
the appeal of their obvious intelligence, border
collies do not really make good pets. They get
bored and frustrated in urban settings without
room to run free.

59

ABOVE TOP & BOTTOM:
Australian kelpie
A worker to its core, this
Australian herding dog
has its roots on sheep and
cattle ranches, where it is
seldom idle.

RIGHT:
Bernese mountain dog
One of the largest breeds
in the world, this Swiss
dog was used as an all-
round worker in alpine
farms, even hauling carts
laden with milk and
cheese.

OPPOSITE:
Smooth collie
This Scottish breed makes
a better pet than the
border collie because it
is more good-natured,
though it is less refined for
working as a herding dog.

LEFT:
Estrela mountain dog
This big breed from the
mountains of Portugal is
a guardian dog that lives
among the herd to defend
it from attack.

BELOW:
Maremma sheepdog
This Italian sheepdog
breed has a thick woolly
coat rivalled only by that
of their charges.

Pyrenean sheepdog
This small herding breed comes from the French Pyrenees. What it lacks in size, the sheepdog makes up for in agility. It is a frequent competitor in dog trials.

Tervueren
Also known as the Belgian shepherd dog, the tervueren is bred to herd livestock but is also big and tough enough to take on a guardian role. Like the similar German breed, it makes a good guard dog or police animal.

Central Asian shepherd dog

Also called the Turkmen wolfhound, this big breed traditionally protects livestock in the steppes of Central Asia. Further developed during the Soviet era, there are now long- and short-haired types.

ABOVE TOP:
Kangal dog
This majestic herding dog
is the national breed of
Turkey. It has
a strong protective
instinct and may attack
strangers if it is not
properly handled.

ABOVE BOTTOM:
Hungarian Kuvasz
This guardian dog is
white and shaggy to
help it blend into a flock
of sheep.

RIGHT:
Friaar dog
This Icelandic sheepdog
breed is sure-footed and
agile, ideal for the rough,
rocky terrain in its
volcanic homeland.

New Zealand sheepdog

Sheepdogs in New Zealand are often a mix of border collie, Rottweiler and German shepherd dog.

Shetland sheepdog
This island breed is a
miniature version of the
rough collie, a similar
herding breed from the
mainland (and most
famous as *Lassie*, the star
of a television show).
The breed is also known
as a Sheltie.

OPPOSITE:
Tornjak
This sturdy Bosnian
sheepdog breed is related
to the mountain breeds
of Romania and Greece.

ABOVE TOP:
Spanish water dog
Bred with a long, shaggy coat that dries quickly, this Spanish working breed has been around for more than 800 years but was only officially recognized in the 1980s.

ABOVE BOTTOM:
Komondor
A shaggy Hungarian sheepdog breed that looks like it is in disguise as a sheep. This probably arrived in Eastern Europe with migrants from Asia around 1,000 years ago.

Australian herder
Also known as the cattle dog, this is a very alert and eager breed. It drives cattle by nipping the heels of the larger animals.

ABOVE:
Briard
This breed is the French
equivalent of the Old
English sheepdog. It has
a shorter coat but a
similar shaggy fringe over
the eyes.

OPPOSITE:
Old English sheepdog
A highly distinctive
breed with eyes covered
by long hairs around
the eyebrows. The thick,
fluffy fur requires a lot
of attention. In recent
years, the breed has fallen
in popularity and is now
listed as endangered.

Lagotto Romagnolo
With a name that means
"lake dog of Romagna",
this small Italian breed
is thought to be the
founding type of all later
water dog breeds.

Truffle hunting
The hidden fungal mass
known as the truffle
is highly prized for its
distinctive musty flavour.
Some varieties fetch well
in advance of £1,000 per
kilogram (2.2 lb). Dogs
of all kinds are trained to
sniff out these valuable
fungi.

LEFT & OPPOSITE:
Alaskan Malamute
This wolfish cold-weather sled dog breed is named after the Native American people who bred it.

BELOW:
Sled dogs
Dogs have been used to haul sleds through rugged icy terrain for around 8,000 years. Today, the sled dog breeds are used mostly for racing and leisure.

81

Racing
A sled team of huskies
heads for the start of
the race. The dogs wear
booties to protect their
paws against sharp ice
and rocks.

Siberian husky
This beautiful dog is the most iconic of the Spitz-type breeds, which hail from the ancient breeds of northern Asia and the Arctic. Huskies are perhaps the nearest domestic dog breed to the wild grey wolf.

OPPOSITE:
Guide dog
A Labrador puppy out training to be a guide dog for a visually impaired person. The dog will keep its owner safe while crossing the road or navigating crowded streets.

ABOVE:
Labrador retriever
Always eager to please, the Labrador was bred as a helpful dog that collected downed birds after a shoot or fish that had fallen from a net. They come in several colours: black, chocolate and yellow, though fox red ones (officially classed as "yellow") are increasingly popular.

Guide dog
While the human dictates
the direction, a guide
dog is there to navigate
obstacles and stop when
danger appears.

ABOVE TOP:
Hunting team
Shooting parties use
different breeds of dogs
during hunts. Hounds
help to find game, the
spaniels are there to go
into undergrowth and
flush out and bring back
birds, and Labrador
retrievers also bring
back birds.

ABOVE BOTTOM & RIGHT:
Soft mouths
Retrievers such as this
spaniel are bred to not
bite down on to killed
birds and to return them
to their masters without
damaging them.

ALL:

Greyhound racing
These fleet-of-foot dogs
were bred as sight hounds
that chased down rabbits
and hares for their
masters. They are now
kept as racing dogs that
chase a mechanical hare
around a sand track.

Record breaker
With long limbs and slender, muscled bodies, the greyhound holds the dog speed record. They run at around 70 kph (45 mph), which is twice as fast as humans.

Neapolitan mastiff
Descendants of fighting
dogs bred to entertain in
Roman arenas, this bulky
breed projects immense
strength as a guard dog.

Dobermann
Bred as a fleet-footed scent
hound for tracking fast-
moving prey, this German
breed now has a reputation
as a particularly menacing
guard dog.

Boxer
Named after the way the dogs prod each other with their front paws, this sturdy breed was developed to catch wild boar.

Defensive instinct
Guard dogs are using a primordial instinct to challenge any outsider that approaches the pack. Before actually attacking, the guard dog will snarl, bark and bare its teeth in an attempt to convince an intruder to turn tail and leave.

LEFT TOP:
Muzzles
Security guard dogs need to be trained to bite and so are muzzled during their obedience training until they have learned when to bite and when not to.

LEFT BOTTOM:
Attack dog
Police use intelligent, large dogs, such as the German shepherd dog, to find and apprehend criminals. Few people are able to outrun a police dog.

Beware of the dog
Sometimes the difference between a guard dog and an attack dog is not altogether clear. It's always best to err on the side of caution.

OPPOSITE:

German shepherd dog

This popular breed was renamed the Alsatian (a region of France) in the UK during the First World War, to distance it from Germany – the enemy at the time. The breed originates in this area, which borders Germany.

ABOVE:

Police dog

Often known as K9 units as a pun on the word "canine", dogs are used by police services the world over as attack dogs, trackers and sniffer dogs.

ALL:

Police dogs

Dogs are trained to use
their strength, speed
and intelligence to
apprehend suspects. Just
the presence of dogs is
often enough to maintain
public order.

Ready for action
This police dog working in the Netherlands is wearing a vest. The vest allows the dog's handler to hold and lift the dog more easily.

LEFT & ABOVE TOP:
Rescue
The latest recruit to the Italian School of Water Rescue Dogs is put through its paces. Once it has been thrown into the water, the dog reaches swimmers in trouble and helps to pull them to shore.

ABOVE BOTTOM:
Search
A German shepherd dog has been called in to find survivors buried by an avalanche. The dog can smell people under the snow and begins to dig down to them.

Disaster zone

A detection dog is being trained to find people trapped in a collapsed building using their senses of smell and hearing.

Security check
A security officer and police dog work together inspecting an aircraft at an airfield, looking for stowaways or secreted packages.

ABOVE (ALL):
Sniffer dog
Drug dogs are trained to recognize the scents of illegal substances. If it smells any of them in this luggage the dog will simply sit down next to the bag and await its human handler.

RIGHT:
Beagles
This compact scent hound breed is an ideal candidate as a sniffer dog. Perhaps the most famous cartoon dog of all, Snoopy, is a beagle.

Explosives hunt
This police dog is searching the streets of central London for traces of explosives in order to avert an attack on a public event.

ABOVE:

Weapons search

A Labrador retriever working for the Bureau of Alcohol, Tobacco and Firearms (ATF). It is searching a stadium for planted weapons and ammunition.

RIGHT:

Bomb squad

A dog and its soldier handler are searching for hidden mines and explosive devices.

Show business

An Afghan hound – the most unruly of breeds – steals the show as Crufts, one of the world's most prestigious dog shows, which is held yearly in England. For some dogs, their job is to be put on show as prime specimens of their breed or pedigree. The prize winners will then become hugely valuable as breeding animals.

Companion Dogs

No matter the breed, a dog can make the perfect companion. Providing unquestioning loyalty and support, a dog is always happy to keep you company. Whether the dog is a wolfhound, terrier or sheepdog, it is a member of the family. That said, there are many breeds that have been developed specifically to live alongside us in our homes.

Companion dogs tend to be small, so they do not get under our feet so much, require less to eat and are easier to pick up to carry and cuddle. As such, most companion dogs will be diminutive versions of more workaday breeds developed for hunting or herding. As well as being bred smaller, companion dogs have features that makes them cuter. Dogs that are born earlier and smaller will retain puppy-like features into adult life. These include floppy ears, short legs, and a big head and eyes. Long silky hair, a hindrance to many working dogs, is an appealing trait for pampered pets. In Europe during the pre-modern era, long-haired lapdogs were not only a welcome companion and symbol of wealth and status, but they also provided some warmth for their owner in cold, draughty homes. Fleas were obviously a problem in this setting – for the humans that is. Hairy companion breeds would attract fleas away from their owners.

An alternative anti-flea strategy is to keep a dog without much hair at all, such as the Mexican hairless dog or Chinese crested dog. But in the end, no matter how they look, dogs are loved for just being there.

OPPOSITE:
Norfolk terrier
A cute cousin of dogged rat catchers, this little dog has floppy ears and a shaggy coat that lies close to the body.

Afghan hound

Despite the name, no one knows where this breed originated. It became associated with Afghanistan because it travelled there along the Silk Road with merchants. The dog is a sight hound breed that once caught hares and goats and saw off attacks from snow leopards and wolves. Today, the hound's long, flowing locks make it a spectacular breed that turns heads – not least because it is perhaps the hardest dog to train.

ABOVE:
Chihuahua
This popular Mexican breed is the smallest in the world, being only about 20 cm (8 in) tall when fully grown. The dogs were developed from an ancient breed thought to be used 1,000 years ago by the Toltec people in ritual sacrifices and as food.

RIGHT:
Bearded collie
More compact than other collie and traditional sheepdog breeds, the bearded collie is now more common as a pet. Nevertheless, its rural heritage means it enjoys spending time in open spaces.

LEFT:
Bull terrier
This dog was originally developed as a fighting dog
by cross-breeding bulldogs with various terrier breeds.
However, given firm leadership, today's bull terriers tend to
be mild-mannered and make good, loyal companion dogs.

ABOVE TOP:
Miniature bull terrier
Growing to about two-thirds of the size of the bull
terrier, this breed is very rare today.

ABOVE BOTTOM:
Boston terrier
Another cross between bulldogs and terriers, this North
American breed was developed as a docile companion,
albeit one that requires frequent exercise.

OPPOSITE:
Cesky terrier
A Czech breed that was developed to be smaller than other terriers so it could fit into narrow burrows in pursuit of prey.

ABOVE & LEFT:
Cavalier King Charles spaniel
This breed is related to the older King Charles spaniel, which is named after a 17th-century king of England who was famed for his love of these dogs. This breed's long ears resemble the wig style of the royalist factions in the English Civil War from which England had recently emerged when Charles II was restored to the throne.

LEFT:
Dachshund
Better known by many as a "sausage dog", this short-legged scent hound was originally bred to hunt badgers inside their setts. They are intelligent, friendly and playful and make ideal companions, even though they can be quite hard to train.

ABOVE:
Lakeland terrier
This small and handsome terrier with a wiry coat is no longer used to chase foxes into burrows. However, the agile, fearless terrier is still prone to chase other animals, no matter their size, so it's best not to keep any other pets in the house.

British bulldog
This small offshoot of mastiff breeds is often used as a symbol of Britishness or Englishness because it is said to resemble a corpulent, jowly man who enjoys good living, and who is stubborn but also ready and able to fight if needed. The bulldog has long been in decline in popularity as a pet because of common health problems.

LEFT TOP:
French bulldog
This breed was developed from toy bulldogs, a small British breed that became popular in France in the 19th century. Like all flat- or squashed-face breeds, they commonly suffer health problems.

LEFT BOTTOM:
English mastiff
This huge and old breed is the "dog of war" referred to in William Shakespeare's *Henry V*. Despite their massive size, English mastiffs are docile and courageous and make excellent family pets.

OPPOSITE:
Japanese spitz
This fluffy little breed
is descended from
the hunting dogs that
originate in northern
Asia and around the edge
of the Arctic. It is prone
to excessive barking but
can be trained to behave.

ABOVE TOP:
Clumber spaniel
This terrier has a
long coat with orange
markings. It is named
after the country
estate of the Duke of
Newcastle. They require
frequent grooming and
tend to drool, but make
good family pets.

ABOVE BOTTOM:
Lhasa Apso
Bred as a watchdog for
Tibetan monasteries,
the flowing locks of
this Asian breed won it
many admirers in Europe
during the early 20th
century.

Leonberger

Named after a Bavarian town, this big breed was developed as a mountain herd dog. The dog (male) has a shaggier and wider head than the bitch (female, left). Submissive, friendly and gentle, they are ideal family pets for those who can accommodate their giant size and cope with the hair!

ABOVE TOP & BOTTOM:

Pekingese
DNA analysis suggests
that this is one of the
oldest distinctive dog
breeds of all: it was
recorded 1,400 years ago
in the Chinese imperial
court. The name refers
to Peking, an old name
for the Chinese capital,
Beijing.

RIGHT:

Staffordshire bull terrier
This tough-looking breed
is actually well suited
to family life. The bull
terrier was developed
from fighting dogs that
were strong and vicious
with other dogs but calm
with humans. It has now
had its fighting streak
bred out but retains the
bruising looks.

OPPOSITE:
Papillon
This old breed means "butterfly" in French and refers to the shape of its ears, which resemble the wings of that insect.

LEFT TOP:
Shih tzu
Originally descended from the "lion dogs" of Tibet, this breed's name means "little lion" in Chinese. It has grown in popularity worldwide over the last half-century.

LEFT BOTTOM:
Shar-Pei
With a name that means "sandy coat" in Chinese , this breed closely resembles fighting dogs shown in pottery artefacts from the Han Dynasty around 2,000 years ago.

Sussex spaniel
This long-bodied breed
is less successful as
a gundog than other
terriers but fits well into
households as a pet.

OPPOSITE:
Scottish terrier
Small but agile, this breed
hails from the Highlands
of Scotland, where it
was developed to hunt
vermin. It is distinctive
for its black hair and
bushy eyebrows.

ABOVE:
Standard poodle
Originally bred as a
water dog, probably in
Germany, for flushing
and retrieving waterfowl,
the standard poodle has
dense, curly fur that keeps
cold water away from the
skin – and is easy to trim
and maintain. The term
"standard" refers to the
size, whereas other poodle
breeds and crossbreeds
tend to be smaller.

ABOVE TOP & RIGHT:

Yorkshire terrier
Small and spirited, the Yorkie is a very popular pet breed. Given good training, the dog makes a great addition to the family, but if neglected it may become noisy and aggressive.

ABOVE BOTTOM:

Maltese
The ancestors of this cute-looking breed were thought to be living around the Mediterranean Sea around the year 300 BC.

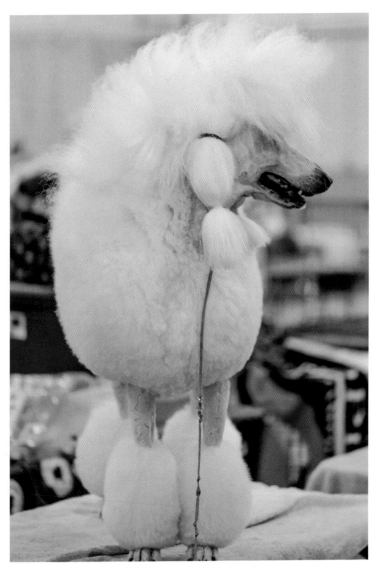

ALL PHOTOGRAPHS:
On show
A standard poodle
(opposite) and Yorkshire
terrier (above) at
the groomers. Good
grooming keeps the skin
and coat healthy – and
looking good. Dogs
raised for showing at
competitions are prime
examples of their breeds.

Puppies

Dogs are ready to breed from as early as six months old, but more likely after the age of one year. Puppies are born after a pregnancy of about two months. Most litters contain three or four puppies but could have as many as ten, or more.

Dogs are altricial, which means they are born in a relative state of helplessness. The puppies are born inside the amniotic sac, and their mother frees them and cleans off the membranous bag, eating much of it. The puppies cannot open their eyes until the age of two weeks, and they are not able to walk straight after birth. However, they are able to smell their mother's milk and fight their way to a teat. Dogs have ten teats, and as a rule of thumb, that is enough to sustain a litter of five pups. Larger litters tend have at least one runt – a weaker member that will probably be muscled off the food supply by its bigger siblings, and needs a helping human hand if it is to survive. The pups learn from their mother how to play safely and not to bite and injure each other.

In the wild, the surviving members of the litter will spend the rest of their lives together as a large family unit. In domestic settings, the litter is generally fragmented at the age of eight to ten weeks. At this age, the puppy is strong enough to walk and play a little with its new family and begins to form life-long bonds with its human owners.

OPPOSITE:
Alaskan Malamute
We are instinctively drawn to protect something as cute as this fluffy pup for the same reasons we look after our own babies.

LEFT:
Bearded collies
Pups learn through play. One of the earliest lessons they learn is to differentiate their family group from outsiders. That journey begins with the puppy's siblings but continues when they join a human family.

ABOVE TOP:
Airedale terriers
A mother shows her pups how to behave around other dogs – say hello and don't bite!

ABOVE BOTTOM:
Chihuahua
These little dogs reach full size at the age of about ten months.

Irish wolfhound

A pair of young wolfhounds exchange a lick. Licking is a signal of submission and trust. You would not lick your enemy!

ALL PHOTOGRAPHS:
Growing fast
Puppies develop quickly after birth. Within a couple of weeks, they start to do dog-like things, such as wagging the tail, barking and generally having fun.

As a puppy is integrated into the home of its new owners, it is a good idea for them to meet a wide range of people from outside the family during the first few months.

ABOVE (ALL):
Bull mastiff
Play is a busy time for these mastiff pups. They are building mobility skills and control and also learning to interact with each other. In large sibling groups, a hierarchy is put in place through play. During training, the dog is given its place in its human family.

Bloodhound mother and puppy
Most puppies are entirely dependent on their mother for the first four weeks and then spend more time away from her. They are weaned around week eight.

Pedigree puppies
Maintaining the pedigree, or quality of breeds, means careful breeding to ensure that puppies do not inherit unwanted or indeed harmful traits. Pictured here are border collies (right), bloodhounds (below) and dachshunds (opposite below).

OPPOSITE TOP:
Dachshund
A sausage dog
puppy relaxes on a
comfortable rug.

OPPOSITE BOTTOM:
Suckling
A litter of seven-day
-old shih tzu puppies
have a meal of their
mother's milk.

LEFT:
Standard poodle
Puppies, including these
young poodles, are
seldom far from their
mother's side for the first
ten weeks of life.

PREVIOUS PAGES:
Puppy farms
Dogs are bred for the
pet market at puppy
farms and breeders. The
puppies spend the early
days of their lives as part
of the gang.

RIGHT:
Afghan pup
It will take several
months for this puppy to
grow the length of locks
shown by its glamorous
mother.

OPPOSITE:
Cocker spaniel
Named after its job of
flushing woodcock and
other ground birds,
this cocker spaniel pup
probably has a more
comfortable life ahead
of it as a pet.

Japanese Akita
A family of Akitas take the opportunity to rest in the warmth. Akita puppies need a long period of training that starts early in order to be a fulfilling pet.

RIGHT:
King Charles spaniel
This little pup is already showing off the long, pendulous ears typical of this breed.

OPPOSITE TOP LEFT:
Bull terrier
Even though it's only young, this bull terrier pup cuts an impressive, muscular, figure.

OPPOSITE TOP RIGHT:
Basenji
As it gets older, this African dog will become chunkier.

OPPOSITE BOTTOM LEFT:
Boston terrier
The puppy's ears grow stronger with age and will be pointed by adulthood.

OPPOSITE BOTTOM RIGHT:
Cesky terrier
This Czech dog needs a lot of combing and trimming to keep its soft coat in good condition.

ALL PHOTOGRAPHS:

Out and about
Once strong enough
– and fully vaccinated
– puppies should be
allowed out and about
to explore. The first 100
days of life are a critical
time for the dog, when it
learns so much about
the world.

Border collie

A pair of border collie puppies plays on the beach. These dogs will probably have little time for play once they are old enough to work as sheepdogs – and that is the way they like it!

LEFT:
Italian greyhound
These racing dogs will
start competing around
the age of
18 months.

ABOVE TOP:
Hungarian kuvasz
This guardian dog is
bred to attack any threat.
Puppies need to be taken
in hand early if they are
to be good pets.

ABOVE BOTTOM:
Komondor
This pup will not develop
the tasselled coat for
which this breed is famed
for at least a year.

ABOVE:
English mastiff
Smell is the main sense used by pups. They are deaf and blind at birth.

RIGHT:
Italian spinone
Puppies are able to swim safely at the age of about ten weeks.

OPPOSITE:
English springer spaniel
As they get older, puppies need to be exercised more frequently and for longer.

OPPOSITE TOP:
Norwegian elkhound
From a young age, this breed is impervious to cold thanks to its thick coat.

OPPOSITE BOTTOM:
Otterhound
Bred for hunting in water, this dog has a naturally oily coat and needs frequent washing and grooming to prevent tangles.

LEFT ABOVE:
Neapolitan mastiff
The loose-fitting skin typical of this breed is apparent from a young age.

LEFT BELOW:
Norfolk terrier
Bred to hunt as part of a pack, this little dog is highly social.

Dog Behaviour

Beneath the different outward appearances of body breeds, deep down, all dogs are wolves. This is partly why dogs makes such great companions. In the wild, wolves live in family groups with a complex social structure that's managed by a language spoken by all dogs – a language of visual signals and smells. A contented pet dog behaves well because it has found its place in our families.

Training techniques harness the animal instincts of a dog and put them to good use. These instincts include the urge to follow a scent and to chase potential prey. The dog also plays its part in protecting the family, dog or human, by challenging intruders.

Despite fitting in well with a human group, the dog is a very different animal. Wild dogs are seldom idle and must cover long distances and work hard to find food. That job has been taken for a pet dog but the urge to move and the need to exercise and interact with its family through play remains. Depending on breed, most dogs require around 60 minutes of walking, running and playing a day – perhaps more.

Wagging tails, barking, growling, sniffs, licks and soft bites are all examples of the ways dogs communicate with each other and they will use these signals just the same in a human group. If trained well, kept stimulated and led with clear instructions, the dog will be happy.

OPPOSITE:
Howl!
This beagle is letting any nearby dogs know where it is. Either that or it has heard a high-pitched sound and mistaken it for a fellow dog's call.

Wag the tail

The dog's tail is an important communicator. A relaxed dog wags its tail in a relaxed way. When it's nervous, the tail is held lower than usual. When the tail is held high, the dog is excited and stimulated by something.

OPPOSITE (BOTH PHOTOGRAPHS):
Smell hello

Dogs live in a world of scent and the first thing they do upon meeting another dog is to smell them. The scent not only identifies them as a friend or stranger but it also contains information about a dog's mood.

Barking

The howl is a wide-area signal, but barking is a more direct form of communication. It could be a call hello, or a question: where have you gone? It could also be an order: stay away or else.

190

Play time
This border collie is busy
having fun. Perhaps more
than any other common
breed, this dog needs
exercise and stimulation.

ALL PHOTOGRAPHS:
Walkies!
Dogs are built to move and need to have the opportunity to walk, run, explore and socialize every day, preferably for at least an hour.

Cool

This Boston terrier dog is heading for a swim. Dogs do not sweat as efficiently as humans and so a cooling dip is hard to resist when the weather is hot.

ABOVE TOP & RIGHT:
Cavalier King Charles spaniel
Impressive for a small dog, this breed can jump more than 3 m (3.3 yards). They certainly enjoy the attention when they perform.

ABOVE BOTTOM:
Dachshund
With only short legs, these dogs avoid deep water, but some take pleasure in a paddle.

BOTH PHOTOGRAPHS:
Serious games
A pet pooch needs its wild instincts tended to, and this corgi (below) and boxer (opposite) like nothing more than chasing a ball. To them, it is just as important as hunting a rat or rabbit.

English springer spaniels
A trio of spaniels
compete with each other
to take the ball back to
their owner.

Different job

Bulldogs were bred for bull-baiting, a sport in which dogs were sent to attack a tethered bull. This activity has been banned for almost 200 years, and so bulldogs have another job: to look cute.

Street dogs
Three young street dogs huddle together when sleeping on the road. When domestic dogs live wild, they become feral and revert to natural behaviours of wild dogs.

ABOVE:
Tricks and training
A pharaoh hound loves
to show off. Dogs enjoy
learning tricks. It keeps
their minds occupied and
earns them some praise.

RIGHT:
Shake it off
A German shepherd dog
shaking off water after
a swim. The long outer
guard hairs prevent much
of the water from getting
to the fluffy underfur.

ABOVE:
Jowls
The long, slobbering jowls of a scent hound keeps the lining of the nose moist and more sensitive to scents in the air.

RIGHT:
Feeding time
Dogs are creatures of habit. They know when it's time to eat and they make sure you do too!

ALL PHOTOGRAPHS:

Eating

Dogs are carnivores and have a short gut that processes nutrient-rich meat in a few hours. Pet dogs can overeat – and be overfed – easily and enjoy dry foods that are less packed with calories. A chew toy or bone will keep the teeth and gums in good condition.

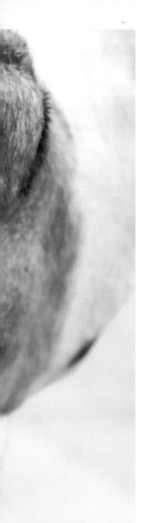

ALL PHOTOGRAPHS:
Listening
Dogs are deaf at birth but
start to pick up sounds
soon after. When fully
developed, a dog's sense
of hearing is four times as
acute as a human's. Most
dogs are able to direct
each ear independently to
pinpoint the location of
a sound.

ALL PHOTOGRAPHS:
Fighting
Conflicts between dogs
are sadly inevitable from
time to time. Some dogs
are not socialized well
and will attack dogs they
perceive as attempting to
dominate them. Fights
are noisy and quick, with
growling and snarls, but
mercifully seldom result
in serious injury.

No fear
This shepherd dog has entered into a dispute with a bull mastiff – one that it cannot win. A dog can't learn bravery but nor can it unlearn it. This kind of behaviour is purely instinctive.

Sociable animal

Many domestic animals – including the sometimes misunderstood and maligned house cat – are highly social animals. Here, this Malamute is perfectly at home with a horse, and visa versa.

Fun, fun, fun
A Leonberger mother
and pup have some fun
together. Dogs are good
at doing that!

Picture Credits